FOSS Science Resources

Solids and Liquids

Full Option Science System
Developed at
The Lawrence Hall of Science,
University of California, Berkeley
Published and distributed by
Delta Education,
a member of the School Specialty Family

1487701
978-1-62571-305-6
Printing 3 — 4/2016
Standard Printing, Canton, OH

Table of Contents

Everything Matters

The world is made up of many things. Trees, bubbles, slides, and drinking fountains are just some of them. These things may all seem very different. But in one way, they are all the same. They are all **matter**. Matter is anything that takes up space.

Matter can be divided into three groups called **states**. They are **solid**, **liquid**, and **gas**.

A slide is a solid.

Water is a liquid.

Bubbles are filled with **air**. Air is a gas.
How did the gas get into the bubbles?

Gases are hard to see and feel. You can't
hold gas in your hand or see it in a bucket.
How can you see gas?

Air is gas, and air is all around. You can see
air make a windmill spin. Which windmills
show air at work?

A helium balloon is fun. Helium is a kind of gas. It is lighter than air. What happens when you let go of a helium balloon? It floats away.

We use solids and liquids all the time. Every solid and liquid is different. So they are useful in different ways.

Cement bricks are strong and hard. They are just right for building walls.

Wool is soft and flexible. It is good for hats and scarves.

Water can spray and splash. It makes
a hot summer day lots of fun!

Look around you for solids and liquids.
How will you use them today?

Solid Objects and Materials

Chairs are solid **objects**.
Blocks are solid objects.
So are chopsticks.

Chairs, blocks, and chopsticks are all different. We sit on chairs. We build with blocks. We eat with chopsticks.

Chairs, blocks, and chopsticks are all the same, too. They are all made of the same **material**. They are all made of wood.

Wood has good **properties** for making chairs.
Wood is strong and rigid.

But wood is not a good material for making socks.
What material has good properties for socks?

Fabric is a good material for socks. Fabric is soft
and flexible. Fabric is a good material for shirts
and blankets, too.

Kick balls are solid objects. Rubber is a good material for kick balls. Rubber stretches, and it is strong.

Rubber is a good material for making tires and balloons, too.

Some shoes are made of fabric, rubber, and metal. Some shoes are made of the material leather. Leather is strong and flexible.

Shoes are solid objects. This shoe is made of three different materials. Can you see all three materials?

Jars are solid objects. This jar is made of two materials. Can you see the two materials?

The jar is made of the material plastic. Plastic is strong and light. The label is made of the material paper. Paper is light and flexible.

Windows are solid objects. Windows are made of the material glass. Glass is strong and **transparent**.

What other objects are made of glass?

Cars are solid objects. Cars are made of many materials. How many different materials can you see?

Thinking about Solid Objects and Materials

1. What material would be good for making an umbrella?

2. How many materials are used to make a pencil?

3. How are materials different from objects?

Towers

This is a **tower**. It is called a communication tower. It is rigid and tall. The base is wider than the top. It can stand by itself. It has a very strong steel frame.

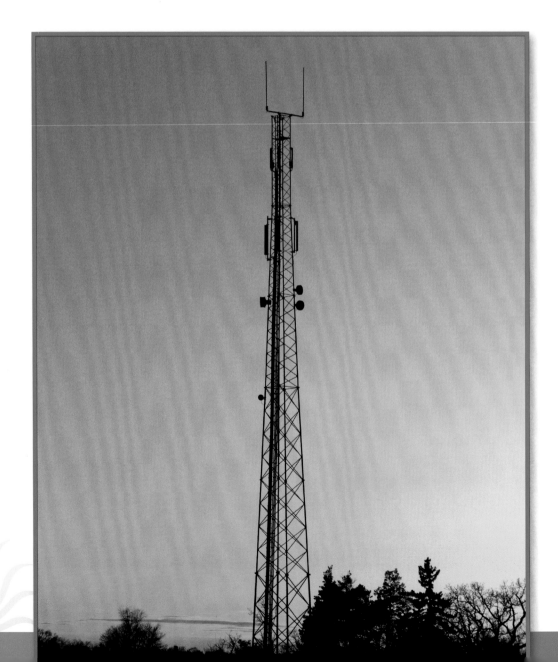

This fire lookout tower is on a mountaintop in the forest. It is a tall structure. Forest rangers use it to look for forest fires. The lookout tower is rigid and has a wide base. A forest ranger can see the whole forest from the top.

One of the most famous towers in the world is the Eiffel Tower. It was built in Paris, France, over 120 years ago. The tower's wide base and narrow body make the Eiffel Tower very stable.

This structure is in Seattle, Washington. It is called the Space Needle. It is tall and rigid. It has a wide, strong, and heavy base. Do you think it is a tower?

Bridges

A **bridge** connects two landmasses. Bridges usually go over water.

The picture shows the Golden Gate Bridge. This bridge carries a road across San Francisco Bay in California.

Two long steel cables hold up the roadway. Two huge towers support the steel cables.

The Tower Bridge connects the two sides of a river in England. It is two bridges in one. The lower bridge is a roadway that rises to let large boats pass. People walk on the upper bridge to cross the river.

Here is a simple bridge in a meadow. What materials is the bridge made from? How is the bridge supported?

A bridge allows you to walk over a creek without getting your feet wet. What are the parts of this bridge?

This bridge helps people walk across the canyon safely. Steel cables hold up the narrow footpath.

Liquids

They can splash. They can squirt. They can bubble and fizz. They come in many colors. What are they? Liquids!

Liquids can spill. Liquids can flow. That's why liquids have to be kept in containers. Glasses, bottles, and tank trucks are containers.

Do you know what the largest
liquid container in the world is?
It's the ocean!

Liquids flow. That's why they can change shape.
These four glasses are the same size and shape.
Each glass has the same amount of liquid.

These containers are different sizes and shapes.
Let's pour liquid from three of the glasses into
the containers.

The water looks different! It is a different shape. The same amount of water can be tall and thin. It can be short and wide.

Liquids change shape in each kind of container.

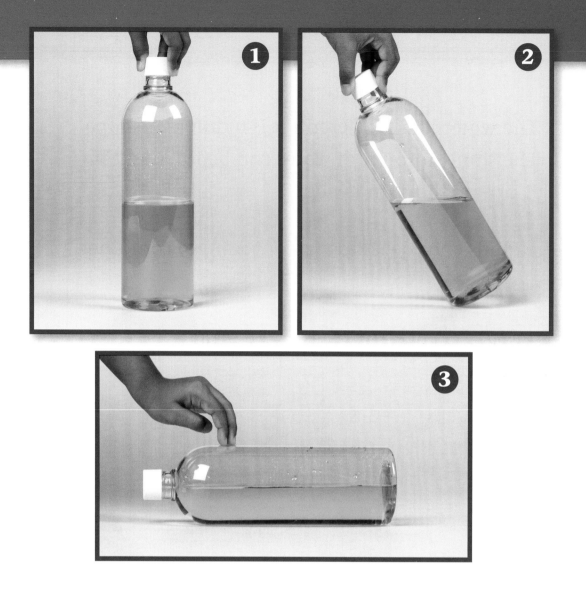

Liquids always move to the bottom of a container.
The liquid is in the bottom of this bottle.

Look at the pictures of the bottle turning over.
Compare pictures 1 and 3. Did the liquid move?
Or did the bottle move? What is different about
the liquid in all the pictures?

What is the same about the liquid in all the pictures? Look at the line near the **surface** of the liquid. It doesn't matter how you turn the bottle. The surface is always flat and level.

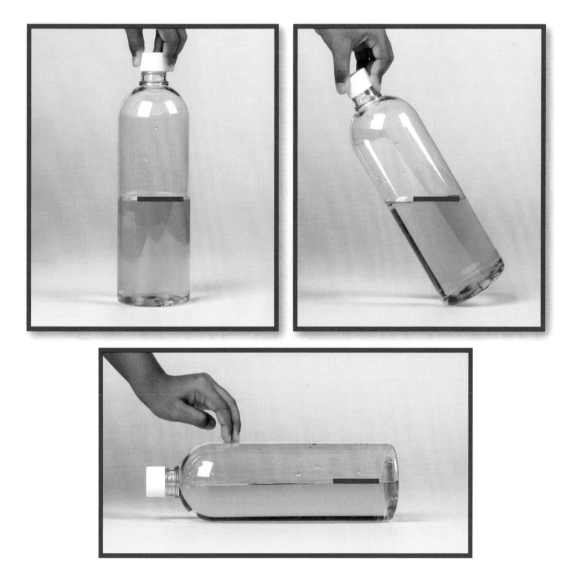

Pouring

Rocks are solid objects. This mountain
is a giant rock. Boulders are big rocks.
River rocks and gravel are smaller rocks
that you can pick up.

A piece of sand is a tiny rock. We call pieces of sand **particles**. You can put millions of sand particles in a bucket.

Sand can pour out of a bucket. Is sand a liquid or a solid?

The surface of a liquid is always flat and level. Is the surface of sand flat? Is the surface of sand level?

Heavy, solid objects **sink** in liquids.
Do heavy objects sink in sand?

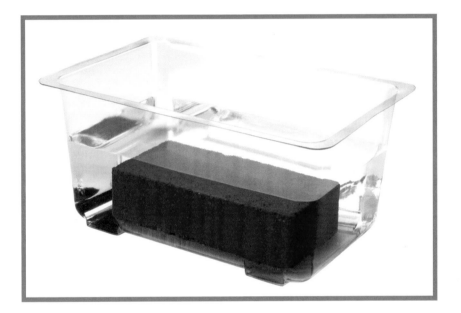

What happens when you pour sand and water on a hard surface? The sand makes a pile. But the water flows and spreads out.

Here are some other materials that pour.
Are they solids or liquids?

Comparing Solids and Liquids

What is the difference between solids and liquids? They have different properties. Properties describe how something looks or feels.

Shape and size are two properties of solid objects. The shape and size don't change unless you do something to the objects. Solids can be rigid, like a bat. When something is rigid, you can't bend it.

Solids can be flexible, like a sweater. When something is flexible, you can bend and stretch it.

Some solids can be broken into pieces. Each piece has a different shape and takes up less space.

What happens when you put the pieces back together? The solid has the same shape as before. It takes up the same space, too.

Solid objects can be very small, like sand.
You can pour sand out of a bucket. But
every grain of sand is a solid.

Liquids have properties, too. A liquid can be poured. It doesn't have its own shape. It takes the shape of the container that holds it.

A liquid has a different shape in each different container.

Liquids can be **foamy**, **bubbly**, or transparent.
They can be **translucent** or **viscous**.

viscous

translucent

foamy

bubbly and transparent

Solids and liquids are all around you. Can you find the solids in each picture? Can you find the liquids?

Mix It Up!

When you put together two or more things, you get a **mixture**.

Can you see what things make up this mixture?

What happens when you mix two liquids?

It depends. Sometimes the liquids mix together to make one new liquid.

Sometimes they don't mix. The two liquids form layers.

What happens when you mix a solid with a liquid?

It depends. Solids like marbles just get wet. A solid like a cookie falls apart. It breaks into smaller pieces.

A solid like salt disappears in a liquid.
It **dissolves**. The salt breaks apart into
tiny pieces. The pieces are so small you
can't see them.

How can you find out if the salt is there? If you
wait long enough, the liquid will **evaporate**.
The water will go into the air. The salt is left
behind in the form of **crystals**.

Solids and liquids are everywhere. We use them every day. How are solids being used here?

How are liquids being used here?

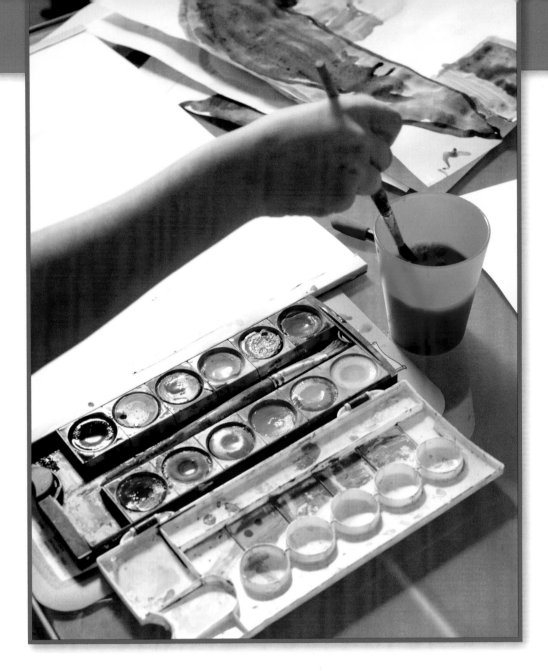

Thinking about Mix It Up!

1. Tell about a mixture of solids.

2. Tell about a mixture of liquids.

3. Tell about a mixture of solids and liquids.

Heating and Cooling

Have you ever had a glass of lemonade on a hot summer day? After you drink the lemonade, ice is left in the glass. After a while, the ice turns to water. Do you know what happened? The ice **melted**.

When a solid melts, it changes from a solid to a liquid.

Other solids can melt, too. Butter is a solid. But if you **heat** the butter, it melts. Solids melt when they get hot.

Liquids can change to solids, too. Do you know how?

Think about making ice cubes. You pour water in a tray. Then, you put it in a cold place. When the water gets very cold, you have solid ice cubes! When a liquid **cools** or **freezes**, it changes to a solid.

Can you think of other liquids that turn to solid?

Liquid chocolate turns to solid as it cools. Liquid chocolate can be poured into molds. When the chocolate cools, it is solid. That's how candy is made into different shapes!

This candle is melting and freezing at the same time. The wax near the flame gets hot. It melts and turns to liquid. Some of the liquid wax burns to make the flame. Some of the liquid wax flows down the side of the candle. When the liquid wax is away from the heat, it cools and freezes back into solid wax.

Thinking about Heating and Cooling

1. Tell about how solids change into liquids.

2. Tell about how liquids change into solids.

3. The ice in the picture above is melting. Why?

Is Change Reversible?

Place an ice cube in a bowl on a table.
The solid ice will melt. It will change
into liquid water. Put the water in a very
cold freezer. The liquid water will freeze.
It will change back into solid ice.

This change between solid and liquid is **reversible**. *Reverse* means to go the other way. Heating can change water from a solid to a liquid. Cooling can change it back to a solid.

What other materials freeze and melt like water? What other changes are reversible?

How do you make pancakes? Mix flour, milk, egg, and a little oil to make batter. Batter is a thick liquid.

Pour some batter on a hot griddle. After a few minutes, the pancake is ready to eat. The hot pancake is a solid object. The liquid batter changed into a solid. Is making pancakes reversible? No, the pancake cannot change back to batter.

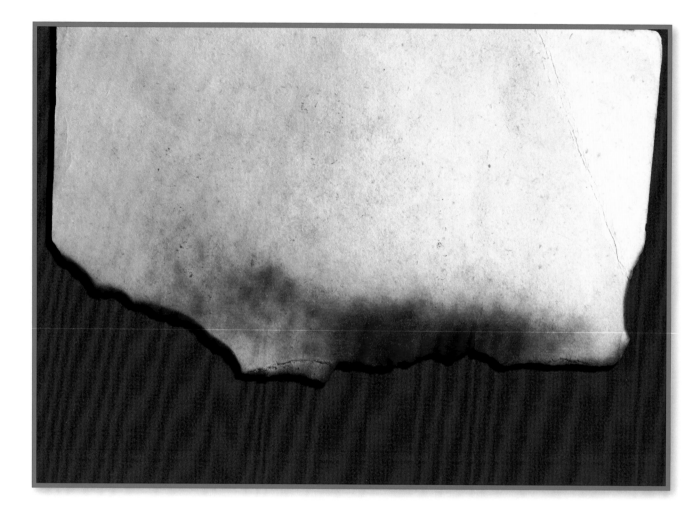

Paper is a solid material. What happens to
paper when you heat it? Does it melt like
ice and turn into liquid? When you put
paper on a hot surface, it changes color.
Paper might turn brown.

With more heat, the paper starts to burn. A little bit of ash remains after the paper burns up. Burning paper is a change that is not reversible.

Did you ever drop a fresh egg? A fresh egg is liquid inside.

Boil an egg in water for a few minutes, and the egg cooks.

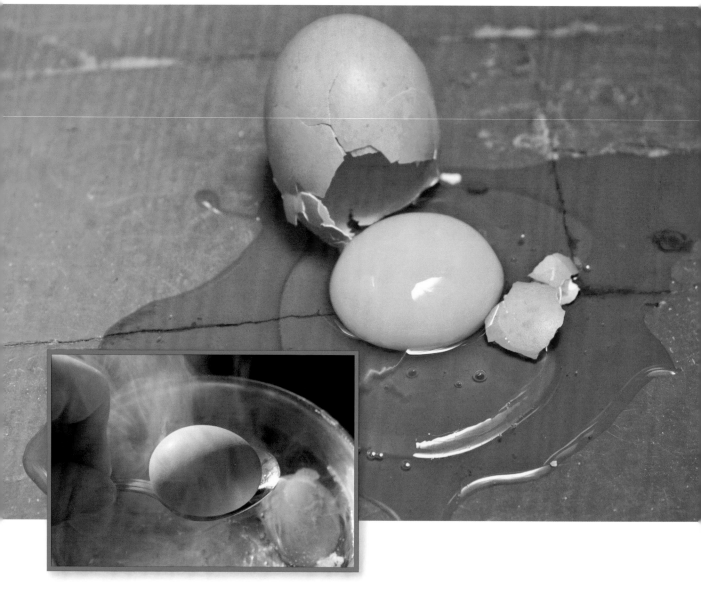

When you crack the shell, the egg inside
is solid. A cooked egg changes from
liquid to solid.

Does cooling the cooked egg change it
back into liquid? No, cooking an egg is
a change that is not reversible.

Thinking about Is Change Reversible?

1. What happens when ice cream warms? Is its state reversible?

2. Chocolate can be a liquid or a solid. How can you reverse its state?

3. Explain how the change in a wax candle is reversible. Explain when the change is not reversible.

4. Describe another example of a change caused by heat that is not reversible.

Glossary

air a mixture of gases that we breathe **(6)**

bridge a structure that connects landmasses over water **(26)**

bubbly describes a liquid that is full of bubbles **(50)**

cool to make something colder **(64)**

crystal the shape of salt after evaporation **(58)**

dissolve when a solid is mixed with a liquid, and the solid breaks apart into pieces so tiny they can't be seen in the liquid **(57)**

evaporate when a liquid dries up, goes into the air as a gas, and can't be seen **(58)**

fabric a flexible material used to make clothing. Fabric and cloth are the same. **(14)**

foamy describes a liquid that has a layer of bubbles on top **(50)**

freeze to change a liquid to a solid by cooling it **(64)**

gas matter that can't be seen but is all around. Air is an example of a gas. **(4)**

heat to make something warmer **(63)**

liquid matter that flows freely and takes the shape of its container **(4)**

material what something is made of **(13)**

matter anything that takes up space **(3)**

melt to change a solid to a liquid by heating it **(62)**

mixture two or more materials put together **(54)**

object a solid thing **(12)**

particle a tiny piece of a material **(39)**

property something that you can observe about an object or a material. Size, color, shape, texture, and smell are properties. **(14)**

reversible to change back to the original state **(69)**

sink to fall or drop to the bottom **(41)**

solid matter that holds its own shape and always takes up the same amount of space **(4)**

state one of the three groups of matter: solid, liquid, or gas **(4)**

surface the top layer of something **(37)**

tower a tall structure **(22)**

translucent describes a liquid or solid that is clear enough to let light through but is not clear enough to see something on the other side **(50)**

transparent describes a liquid or solid that you can see through easily **(19)**

viscous describes a liquid that is thick and slow moving **(50)**